Jill wants to ride a cable car.

"Hop on, Jill!" says Dad.

1

"This cable car is full!"
Jill says.

Dad says, "It goes to the Nibble Shop. We can get off there."

"At times cable cars need a push!" Dad says.

"What makes them work?" asks Jill.

"We'll sit next to this bush," says Dad. Then he puts a scribble pad on the table.

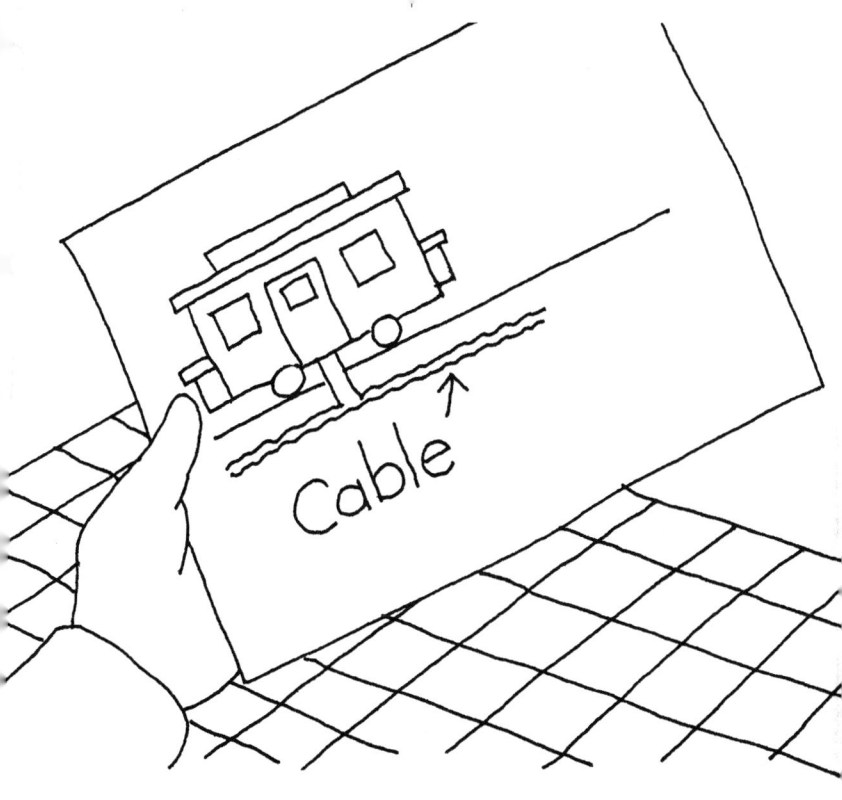

Dad tells Jill, "A wire cable runs under the street. The car grips the cable. The cable pulls the car. Then the car goes."

"A man ran the car we were on," says Dad. "He makes the car grab the cable."

"Where does a cable start?" Jill asks.

"Cables start here. They get the spark they need to run."

"I have been on a cable car. I hope we can help push one!"

The End

Understanding the Story

Questions are to be read aloud by a teacher or parent.

1. What does Jill want to do first?

2. What does Dad do with his scribble pad?

3. Where does a cable wire go?

4. What is Jill's cable car wish?

Answers: 1. ride a cable car 2. Possible answer: draws pictures to explain cable cars 3. under the street 4. to help push one cable car up

---

© Saxon Publishers, Inc., and Lorna Simmons

All rights reserved. No part of the material protected by this copyright may be reproduced or utilized in any form or by any means, in whole or in part, without permission in writing from the copyright owner. Requests for permission should be mailed to: Copyright Permissions, Harcourt Achieve Inc., P.O. Box 27010, Austin, Texas 78755.

Published by Harcourt Achieve Inc.

Saxon is a trademark of Harcourt Achieve Inc.

Printed in the United States of America
ISBN: 1-59141-021-5

1 2 3 4 5 446 09 08 07 06 05

Phonetic Concepts Practiced

-ble (cable)

Nondecodable Sight Words Introduced

been	pull
bush	push
does	put
full	says
goes	wants

ISBN 1-59141-021-5

Grade 2, Decodable Reader 7
First used in Lesson 42

Mr. Fudd's Unlikely Morning

written by Leya Roberts
illustrated by Mernie Gallagher-Cole

THIS BOOK IS THE PROPERTY OF:

STATE_____	Book No. _____
PROVINCE_____	Enter information
COUNTY_____	in spaces
PARISH_____	to the left as
SCHOOL DISTRICT_____	instructed
OTHER_____	

| ISSUED TO | Year Used | CONDITION ||
		ISSUED	RETURNED

PUPILS to whom this textbook is issued must not write on any page or mark any part of it in any way, consumable textbooks excepted.

1. Teachers should see that the pupil's name is clearly written in ink in the spaces above in every book issued.
2. The following terms should be used in recording the condition of the book: New; Good; Fair; Poor; Bad.